High Interest
GEOMETRY

Gunter Schymkiw

World Teachers Press®

Published with the permission of R.I.C. Publications Pty. Ltd.

Copyright © 1999 by Didax, Inc., Rowley, MA 01969. All rights reserved.

First published by R.I.C. Publications Pty. Ltd., Perth, Western Australia.

Limited reproduction permission: The publisher grants permission to individual teachers who have purchased this book to reproduce the blackline masters as needed for use with their own students. Reproduction for an entire school or school district or for commercial use is prohibited.

Printed in the United States of America.

Order Number 2-5106
ISBN 1-58324-028-4

A B C D E F 03 02 01 00 99

395 Main Street
Rowley, MA 01969
www.worldteacherspress.com

Foreword

High Interest Geometry is the fourth book in the popular *High Interest* series. Other titles are *High Interest Language*, *High Interest Vocabulary* and *High Interest Mathematics*.

The goal of this series is to provide students with stimulating activities which consolidate essential skills across a number of curriculum areas.

The activities in this book lead students to explore the many characteristics of, and relationships between shapes.

Clear instructions written in appropriate language guide students in this discovery.

As well as gaining insight into geometric concepts, students will be given opportunities to practice and improve their skills in manipulating, ruling lines, cutting, pasting, folding and using a compass.

Activities are suitable for students in grades five through eight.

As students discover the complexities and patterns revealed in this study, it is hoped that their sense of wonder and desire to find out more is stimulated.

The publisher has chosen to use metric measurements for most activities in this book. The National Council of Teachers of Math supports the use of the metric system as an integral part of the mathematics curriculum at all levels of education (NCTM *Position Statement on Metrication, 1986*). In some activities Imperial (English) measurements are used for illustrative purposes.

Table of Contents

Activity	Page
Straight Lines	6
Curved Lines	7
Straight Line Pattern	8
Rotating Squares	9
Parallel Lines	10
Parallel Lines Again	11
The Pentagon	12
The Hexagon	13
The Octagon	14
The Nonagon	15
The Decagon	16
The Dodecagon	17
Bunch of Squares!	18
Straight Curves	19
Numbers and Letters	20
Tessellation	21
Compass Fun One	22
Compass Fun Two	23
Compass Fun Three	24
Construct-a-Shield	25

Activity	Page
Geometry and Heraldry	26
Instant 360° Protractor	27
Instant Protractor Fun	28
More Instant Protractor Fun	29
Tangram Fun	30
More Tangram Fun	31
Tessellation Again	32
Flower Power	33
Compass Cross	34
Morning Star	35
Concentric Circles	36
Parts of a Circle	37
How Many Squares?	38
Four Cubes	39
Octagon and Rectangle	40
Dividing Shapes	41
Lines and Crosses	42
Cube Models	43
Cut and Make	44
Answers	45–47

Teachers Notes

Introduction

Enrichment and extension are essential components of all learning programs, no matter what the capabilities of the individual student. Providing students with an opportunity to apply knowledge and take understanding to another level is vital. This importance is related to the learning development of the individual as well as to the development of positive learning attitudes. If students can see that they can apply knowledge to solve problems, draw conclusions, identify problems and to extend their existing level of knowledge, then positive attitudes develop that will enhance further progress.

The *High Interest* series of books aim to provide the opportunity for individuals to challenge and extend themselves. This title *High Interest Geometry* focuses on an area of math that is challenging in many forms and involves the use of a range of manipulatives, as well as problem solving skills.

Strategies

High Interest Geometry can be used as:

1. *Class work*
   Once a concept has been presented in class, the appropriate assignment can be photocopied and completed in class to consolidate and reinforce and extend the concept.

2. *Review of class work*
   At the end of a unit of work, the assignments can be used to assess the student's understanding of a particular concept. This allows you to focus further instruction at the point of need for individual students. This approach also provides a straightforward approach to evaluation and recording of the student's understanding of math.

3. *Enrichment and Extension*
   The activities in this book provide excellent resources to extend the learning of those students who have an existing knowledge of concepts. They are ideal for math learning centers.

4. *Homework activities*
   Each assignment can be photocopied and sent home for students to complete independently over the course of the week or pre-designated time period. Parents/Guardians can assist the student if they are having difficulties. The following approaches are encouraged:
   (a) Assist the student with the process involved without solving the problem for the student, and
   (b) encourage the student to try to solve the problem. Any problems encountered at home should be discussed with you at the earliest possible convenience.

Teachers Notes

Instructions

The instructions provided on each assignment are clear and concise. Each instruction has been carefully written to avoid ambiguity. This allows students to work as independently as they have no need to clarify the question.

Benefits

The benefits of *High Interest Geometry* are many:

1. You can readily evaluate where each student is having success or difficulties.

2. Provides parents with the opportunity to observe how their child is achieving in the area of math, as well as an opportunity to join in with their child in completing the exercises.

3. Opportunities are provided for students to practice, extend and review various concepts treated in class.

4. Students are able to take some responsibility for their own learning.

Conclusion

High Interest Geometry is a useful tool for developing the knowledge and understanding of a broad range of geometry concepts. Students can develop a high level of confidence with the opportunities they are given to consolidate and extend what they learn. Confidence in students leads to success and a positive self-image of themselves as learners.

Straight Lines

In geometry you will learn about **shapes** and **lines**. There are two main types of **lines**: **straight lines** and **curved lines**.

Draw a picture made up entirely of **straight lines**. Use a ruler. Use colors to make it look attractive. Remember, everything has to be straight.

Example:

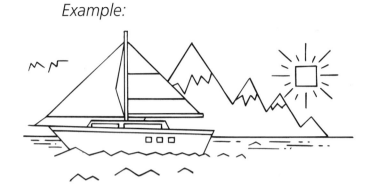

Using a Ruler

Here are some things to remember when you use a ruler:

1. Hold the ruler in the middle. If you hold it at the ends it will tend to slip as you use it.

2. Pull the pencil, don't push it.

Curved Lines

In geometry you will learn about lines. There are **straight lines** and **curved lines**. Draw and color a picture made up entirely of **curved lines**. Use the whole page.

Example:

Color these shapes different colors, cut them out, then reassemble them to make a circle.

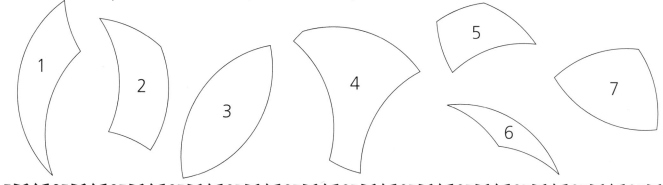

Straight Line Pattern

Use your ruler to draw a straight line pattern. Use black and two other colors to color the pattern in an interesting way.

Example:

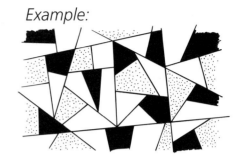

Color these shapes different colors, cut them out, then reassemble them to make a square. There is more than one way of doing this.

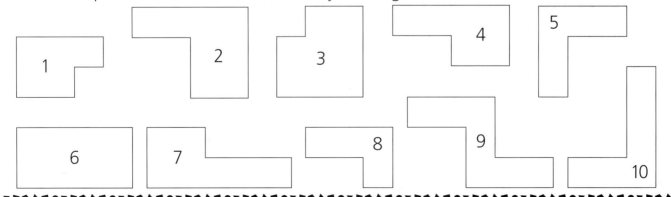

Rotating Squares

Choose three different colors to rule each group of lines.

Color #1 _____   Rule these lines: AD, DG, GJ, JA

Color #2 _____   Rule these lines: BE, EH, HK, KB

Color #3 _____   Rule these lines: CF, FI, IL, LC

Color the rotating squares pattern in an interesting way.

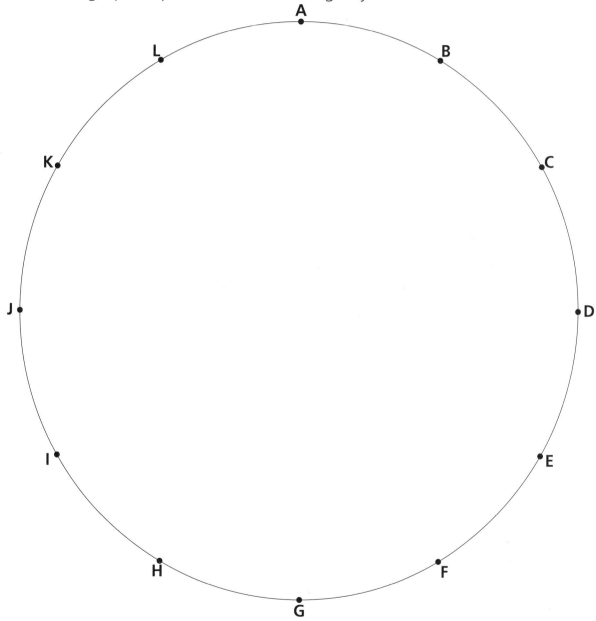

Rule over the lines on the inside to highlight another shape made by this pattern. This shape is called a dodecagon. Write the dictionary meaning of dodecagon.

_____

Parallel Lines

Parallel lines will never meet.

Rule lines AB and CD. Are the lines straight? _____

Now join 1 to 2, 3 to 4, 5 to 6, 7 to 8, 9 to 10, 11 to 12, 13 to 14, 15 to 16, etc.

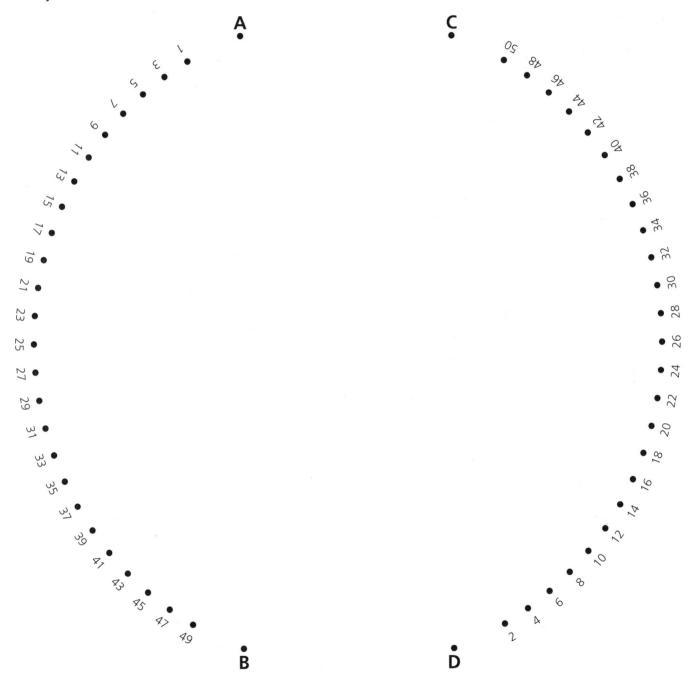

What happened to lines AB and CD?

They appear to _____ . We call this an optical illusion.

Parallel Lines Again

Rule lines AB and CD.

Rule lines beginning at X to each dot in the middle of the page.

Rule lines beginning at Y to each dot in the middle of the page.

What appears to happen to lines AB and CD? _____

The Pentagon

Rule lines: AB, BC, CD, DE, EA

Rule diagonals: AC, AD, BD, BE, CE

A pentagon has _____ sides and _____ corners.

It has _____ diagonals.

Draw a bug in each corner (angle). Color the shape inside the pentagon.

A

E•   •B

D•   •C

This is the Pentagon building in the United States. It is one of the largest office buildings in the world and stretches ≈ 1.6 km around. Approximately how long would each side of the building be?

_____

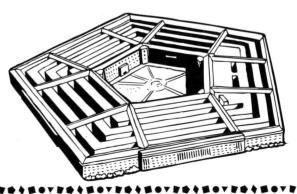

The Hexagon

Rule lines: AB, BC, CD, DE, EF, FA

Rule diagonals: AC, AD, AE, BD, BE, BF, CE, CF, DF

Hexagon Island has _____ sides and _____ corners. It has _____ diagonals.

• A

F •   • B

E •   • C

• D

Jim lives at A. Is this in the north, south, east, or west of Hexagon Island? _____

Color and cut out your hexagon.
Arrange your hexagon with others to form an interesting pattern.

The Octagon

Rule lines: AB, BC, CD, DE, EF, FG, GH, HA

Rule diagonals:
AC, AD, AE, AF, AG, BD, BE, BF, BG, BH, CE, CF, CG, CH, DF, DG, DH, EG, EH, FH

Octagon Island has _____ sides. It has _____ corners.

It has _____ diagonals.

Which direction do you travel going:

1. from A to E? _____
2. from G to C? _____
3. from E to A? _____
4. from C to G? _____
5. from F to B? _____
6. from D to H? _____
7. from B to F? _____
8. from H to D? _____

The Nonagon

Rule lines: AB, BC, CD, DE, EF, FG, GH, HI, IA

Rule diagonals:
AC, AD, AE, AF, AG, AH, BD, BE, BF, BG, BH, BI, CE,
CF, CG, CH, CI, DF, DG, DH, DI, EG, EH, EI, FH, FI, GI

A

I • • B

H • • C

G • • D

F E

A nonagon has _____ sides and _____ corners. It has _____ diagonals.

"Non" means "nine." Write the meaning of:

nonagenarian _____

_____

nonet _____

_____

The Decagon

Rule lines: AB, BC, CD, DE, EF, EG, GH, HI, IJ, JA

Rule diagonals:
AC, AD, AE, AF, AG, AH, AI, BD, BE, BF, BG, BH, BI, BJ, CE, CF, CG, CH, CI, CJ, DF, DG, DH, DI, DJ, EG, EH, EI, EJ, FH, FI, FJ, GI, GJ, HJ

A
J•   •B
I•   •C
H•   •D
G•   •E
F

A decagon has _____ sides and _____ corners. It has _____ diagonals.

"**Dec**" means "**ten**." Write the meanings for the following words:

decade _____

decathlon _____

Decalogue _____

decimal _____

December _____

decimate _____

The Dodecagon

Rule lines: AB, BC, CD, DE, EF, FG, GH, HI, IJ, JK, KL, LA

Rule diagonals:
AC, AD, AE, AF, AG, AH, AI, AJ, AK, BD, BE, BF, BG, BH, BI, BJ, BK, BL,
CE, CF, CG, CH, CI, CJ, CK, CL, DF, DG, DH, DI, DJ, DK, DL, EG, EH, EI,
EJ, EK, EL, FH, FI, FJ, FK, FL, GI, GJ, GK, GL, HJ, HK, HL, IK, IL, JL

A dodecagon has _____ sides and _____ corners. It has _____ diagonals.

Bunch of Squares!

Rule lines: AC, CE, EG, GA, BD, DF, FH, HB, IK, KM, MO, OI, JL, LN, NP, PJ, QR, RS, ST, TQ

Color the pattern using two different colors.

1. Side AC is _____ cm long.
2. Side CE is _____ cm long.
3. Side EG is _____ cm long.
4. Side GA is _____ cm long.
5. What do we call the shape containing the pattern? _____
6. (a) Side GE is _____ cm long. (b) Side OM is _____ cm long.

   (c) Side TS is _____ cm long.

   What is the relationship between the lengths of these sides?

   _____

   _____

Straight Curves

Rule lines: A9, B8, C7, D6, E5, F4, G3, H2, I1
1J, 2K, 3L, 4M, 5N, 6O, 7P, 8Q, 9R
J18, K17, L16, M15, N14, O13, P12, Q11, R10
I18, H17, G16, F15, E14, D13, C12, B11, A10

You can make up many patterns this way. Here are a few for you to consider.

Try these then make up some of your own.

Numbers and Letters

Beginning at "ape," join the dots in alphabetical order. Use your ruler.

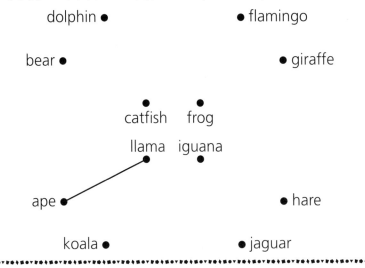

Rule lines to join the numbers and words.

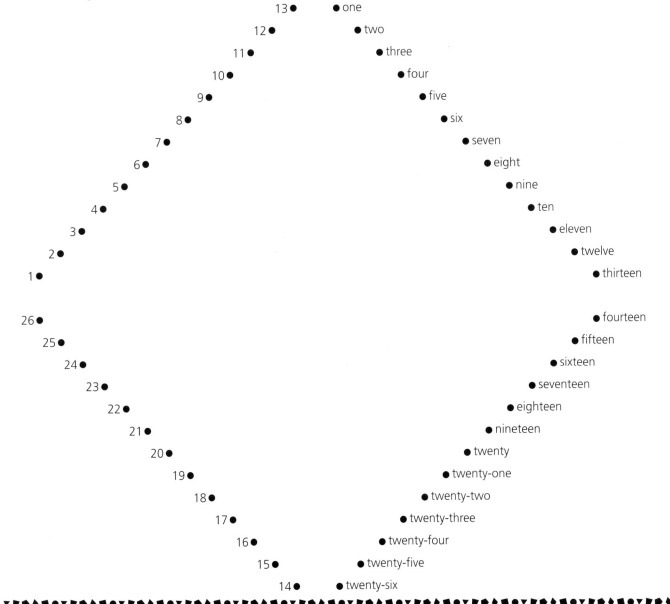

Tessellation

*Shapes **tessellate** if they fit together without spaces. Use your ruler to add more hexagons. Color them to make an interesting pattern.*

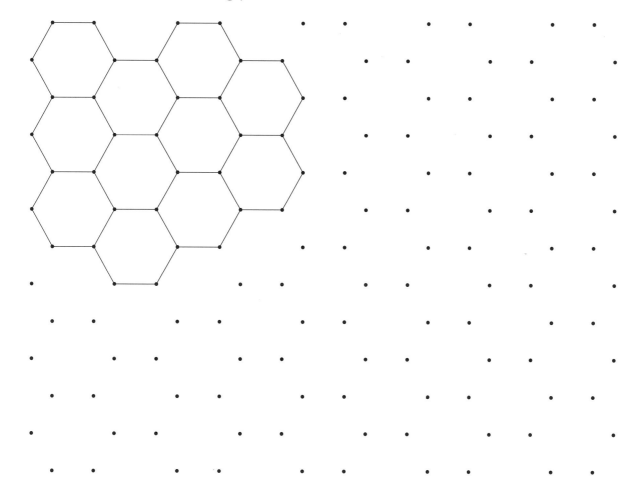

Bee cells are this shape. Bees store honey in them.

Tessellations can be made using one shape or a number of shapes in combinations.

For example, octagons do not tessellate…

but a combination of octagons and squares will tessellate.

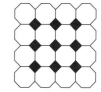

World Teachers Press®   High Interest Geometry 21

Compass Fun One

1. Set your compass at a width of AB.
2. Using A as the center, construct a circle with this width.
3. Keep this width to construct circles with centers at B, C, D, E, F, G, H, I, J, K, L and M.
4. Use two colors to color the pattern in an interesting way.

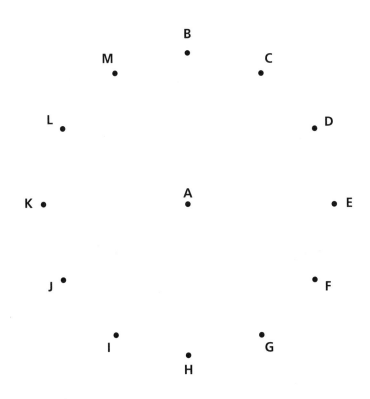

Compass Fun Two

1. Set your compass to a width of AB.
2. Using A as the center, construct a circle using this compass width.
3. Set your compass at the widths below.
   BX, CX, DX, EX, FX, GX, HX, IX, JX, KX, LX, MX
4. Construct circles using these widths. In each case put the point of the compass on the first dot mentioned.

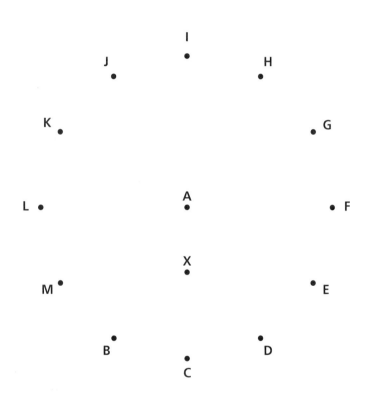

Many English words come from the Latin language. Latin was spoken in ancient Rome. "Circum" in Latin means "around." Write what you think these words have to do with "around."

circus _____

your blood **circulates** _____

Compass Fun Three

Set your compass at the widths below. Construct circles using these widths. In each case put the point of the compass on the first dot mentioned.

AB, CD, EF, GH, IJ, KL, MN, OP, QR, ST, UV, WX, YZ

Make up as many words as you can from the letters in **ellipse**. Fifteen is a good score.

_____   _____
_____   _____
_____   _____
_____   _____
_____   _____
_____   _____
_____   _____

Construct-a-Shield

In the 12th century knights competed in jousting competitions. Jousting knights would charge at each other on horseback. Each knight would try to knock the other from his horse with a long pole called a lance.

For protection they wore armor and carried a shield. As it was difficult to identify riders in armor, special patterns were painted on these shields so spectators could recognize competitors in the jousts.

1. Rule these lines: AB, AC, BD.

2. Set your compass at a width of CD.

3. Construct two arcs using this width.
   Begin with the compass point on C.
   Do the second with the point at D.
   The arcs should cross at X.

4. Draw and color a pattern on your shield.

A•   •B

C•   •D

                        X
                        •

Geometry and Heraldry

Heraldry is the study of crests and shields.

In the 12th century families of knights painted a "coat of arms" on their shields to identify themselves. These usually included a pattern, a picture and sometimes a motto.

1. Rule these lines: AC, BD.
2. Set your compass at a width of XC.
3. Construct a semicircle from X that goes from C to D.
4. Set your compass at a width of AE.
5. Put the point on E and construct arc AY.
6. Put the point on F and construct arc YB.
7. Make up a family coat of arms for your shield.

Instant 360° Protractor

Cut around the outside of the circle to make your own 360° protractor.

Cut out the small diamond shape by lightly folding along the line. This will allow you to check that you are on center when marking an angle.

Keep your instant protractor in a safe place so you can use it again and again. You will have to estimate angles that do not end in a zero or a five.

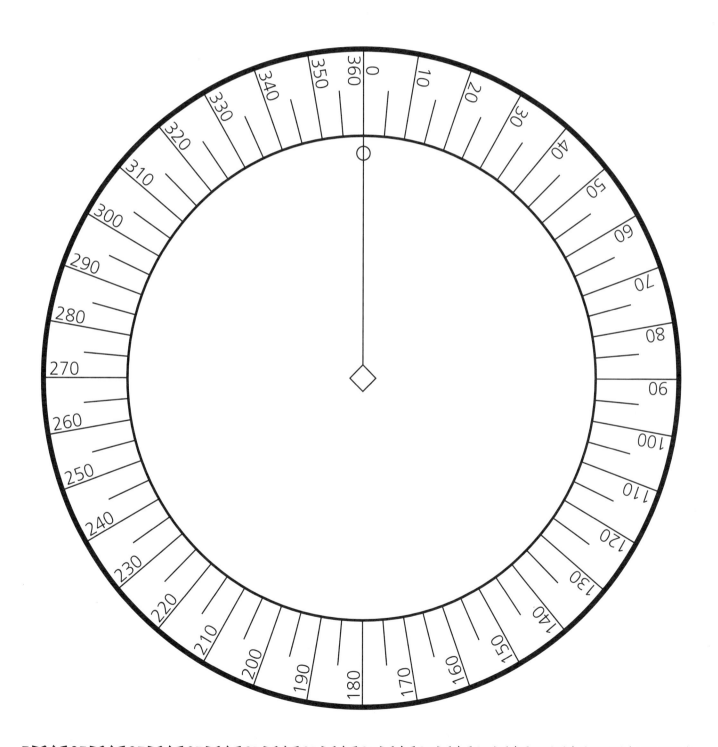

Instant Protractor Fun

Make these angles with your "instant protractor" on a separate piece of paper. In each instance, mark a dot in the center of the diamond to begin with, and then at 0°. Follow the circle around until you come to the angle size required and mark another dot. Rule a line from the center dot to each of the others.

1. Mark and rule an angle of 40°. This is called an **acute** (or sharp) **angle**. Any angle less than 90° is an **acute angle**. Mark and rule three more acute angles. Write their size on the inside as shown.

2. Mark and rule a 90° angle. This angle is known as a **right angle**. It was the "right" angle to use in building. Walls built at angles that were not "right" fell over more easily. It is usually marked with a little square in the corner.

3. Mark and rule an angle of 125°. This angle is known as an **obtuse angle**. Any angle between 91° and 179° is an **obtuse angle**. Mark and rule three more obtuse angles. Write their sizes.

4. Mark and rule an angle of 180°. This is called a **straight angle** for obvious reasons.

5. Mark and rule an angle of 225°. This angle is known as a **reflex angle**. Any angle above 180° is a **reflex angle**. Mark and rule three more reflex angles. Write their sizes.

6. A complete revolution passes through 360°. Have you ever heard someone say, "I was doing 360's on my bike"? What did they mean?

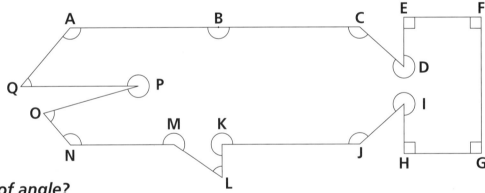

What sort of angle?
Choose from the angle names above to name the angle at:

A _____ J _____

B _____ K _____

C _____ L _____

D _____ M _____

E _____ N _____

F _____ O _____

G _____ P _____

H _____ Q _____

I _____ Color Wally Whale.

More Instant Protractor Fun

It's easy to construct polygons with your instant protractor. Put a dot on your paper just above the angle size given. Label the dots with the letters given then join the dots. Finally, rule all the diagonals given to reveal some interesting patterns.

Pentagon:
A = 72°, B = 144°, C = 216°, D = 288°, E = 360°
Diagonals: AC, AD, BD, BE, CE

Hexagon:
A = 60°, B = 120°, C = 180°, D = 240°, E = 300°, F = 360°
Diagonals: AC, AD, AE, BD, BE, BF, CE, CF, DF

Heptagon:
A = 51°, B = 103°, C = 154°, D = 206°, E = 257°, F = 309°, G = 360°
Diagonals: AC, AD, AE, AF, BD, BE, BF, BG, CE, CF, CG, DF, DG, EG

Octagon:
A = 45°, B = 90°, C = 135°, D = 180°, E = 225°, F = 270°, G = 315°, H = 360°
Diagonals: AC, AD, AE, AF, AG, BD, BE, BF, BG, BH, CE, CF, CG, CH, DF, DG, DH, EG, EH, FH

Nonagon:
A = 40°, B = 80°, C = 120°, D = 160°, E = 200°, F = 240°, G = 280°, H = 320°, I = 360°
Diagonals: AC, AD, AE, AF, AG, AH, BD, BE, BF, BG, BH, BI, CE, CF, CG, CH, CI, DF, DG, DH, DI, EG, EH, EI, FH, FI, GI

Decagon:
A = 36°, B = 72°, C = 108°, D = 144°, E = 180°, F = 216°, G = 252°, H = 288°, I = 324°, J = 360°
Diagonals: AC, AD, AE, AF, AG, AH, AI, BD, BE, BF, BG, BH, BI, BJ, CE, CF, CG, CH, CI, CJ, DF, DG, DH, DI, DJ, EG, EH, EI, EJ, FH, FI, FJ, GI, GJ, HJ

Dodecagon:
A = 30°, B = 60°, C = 90°, D = 120°, E = 150°, F = 180°, G = 210°, H = 240°, I = 270°, J = 300°, K = 330°, L = 360°
Diagonals: AC, AD, AE, AF, AG, AH, AI, AJ, AK, BD, BE, BF, BG, BH, BI, BJ, BK, BL, CE, CF, CG, CH, CI, CJ, CK, CL, DF, DG, DH, DI, DJ, DK, DL, EG, EH, EI, EJ, EK, EL, FH, FI, FJ, FK, FL, GI, GJ, GK, GL, HJ, HK, HL, IK, IL, JL

Use the following table to find the number of diagonals each shape has.

sides	multiplied by =	number of diagonals
4	$\frac{1}{2}$	2
5	1	5
6	$1\frac{1}{2}$	9
7	2	14
8		

sides	multiplied by =	number of diagonals
9		
10		
11		
12		

Tangram Fun

Tangrams are old Chinese puzzles. The seven pieces can be arranged into many shapes.

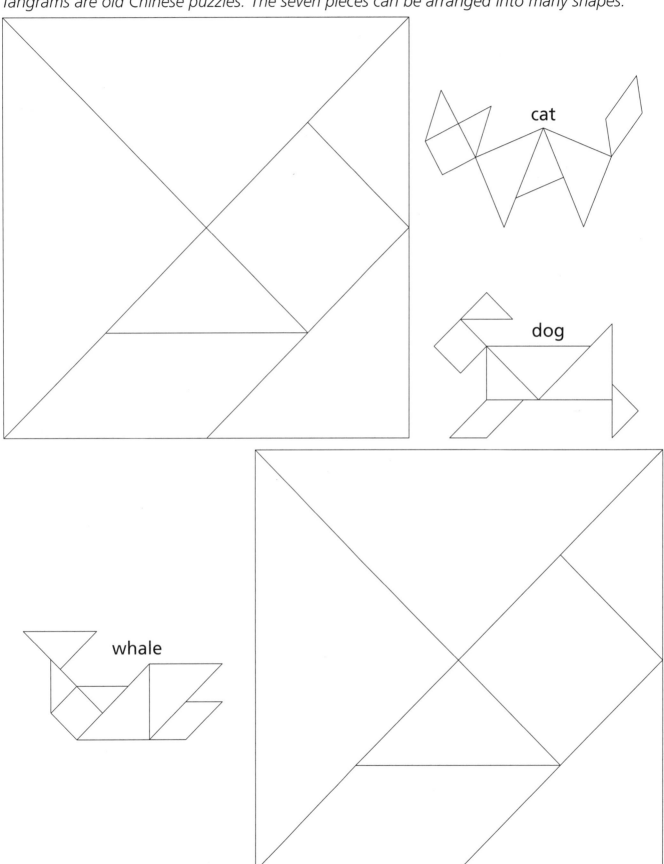

More Tangram Fun

Use the tangrams on page 30 to make two of the shapes below. Draw or paint a picture and paste your tangrams onto it.

For example,

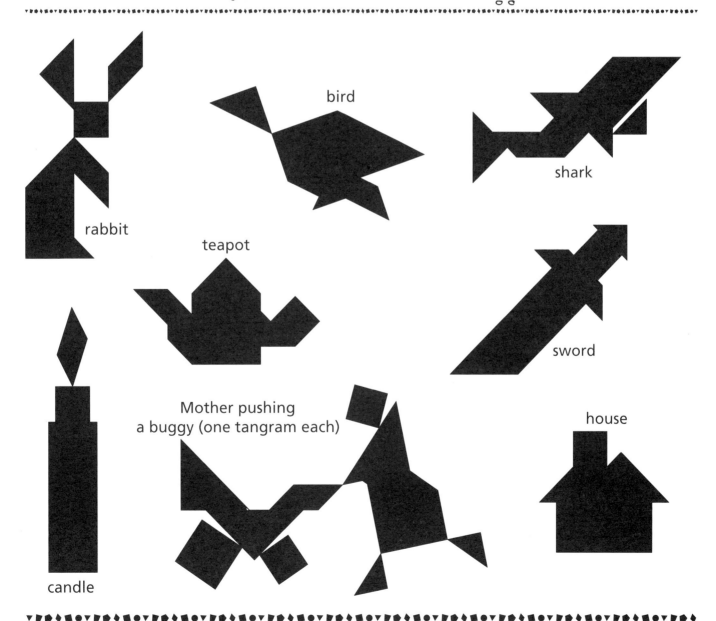

rabbit

bird

shark

teapot

sword

candle

Mother pushing a buggy (one tangram each)

house

Tessellation Again

*Shapes **tessellate** if they fit together without spaces.*
Use your ruler to add more crosses. Color them different colors.

Flower Power

Set your compass width to a distance of AB. Keep this width for all of this activity.

1. Put the point on B and construct arc CD.
2. Put the point on C and construct arc EB.
3. Put the point on E and construct arc FC.
4. Put the point on F and construct arc GE.
5. Put the point on G and construct arc DF.
6. Put the point on D and construct arc GB.

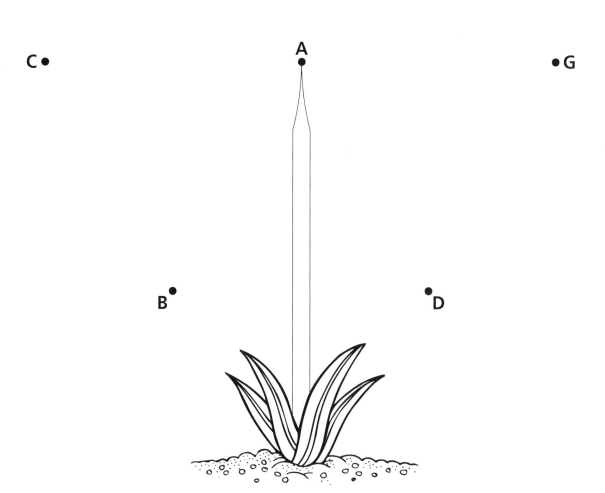

Color your shape an interesting color.
Color the stalk and grass green. Draw a bee visiting your shape.

Compass Cross

1. Rule the lines: AB, BC, CD, DA.
2. Set your compass to a width of EC.
3. Put the point on C and construct arc KH.
4. Put the point on D and construct arc MJ.
5. Put the point on A and construct arc LG.
6. Put the point on B and construct arc FI.
7. Color regions W, X, Y and Z.

1. Rule the lines: AB, BC, CD, DA.
2. Set your compass to a width of EC.
3. Put the point on C and construct arc KH.
4. Put the point on D and construct arc MJ.
5. Put the point on A and construct arc LG.
6. Put the point on B and construct arc FI.
7. Color regions S, T, U and V.

High Interest Geometry 34 World Teachers Press®

Morning Star

1. Rule these lines: AB, BC, CD, DA.
2. Set your compass at a width of AB.
3. Put the point on A and construct arc BD.
4. Put the point on B and construct arc CA.
5. Put the point on C and construct arc DB.
6. Put the point on D and construct arc AC.
7. Rule these lines: EF, FH, HG, GE.
8. Set your compass at a width of EF.
9. Construct arcs FG, HE, GF and EH as in (*3*) to (*6*).
10. Color the pattern in an interesting way.

A

E F

D B

G H

C

World Teachers Press® High Interest Geometry 35

Concentric Circles

1. Circles with the same center are called **concentric circles**.

2. Set your compass at these widths and construct circles with their center at A: AB, AC, AD, AE, AF

3. Join these dots using a ruler and pencil: FG, ML, IH, JK

4. Use two colors to shade your concentric circles pattern.

Parts of a Circle

Set your compass at the width of AB and construct circles 1, 2 and 3 with centers on B.

Circle 1
1. Write circumference on the curved broken line putting one letter to each part.
2. Rule line ABC and label it "diameter."
3. Color the region under the diameter and label it "semicircle."
4. Rule line BD. Color the region bounded by ABD and label it "quadrant."

Circle 2
1. Rule line AB. Label it "radius."
2. Rule line XY and label it "chord."
3. Color the region above the chord. Label it "segment" on the broken line.

Circle 3
1. Rule lines AB and EB.
2. Color the region bounded by ABE and label it "sector."

Circle 1

A •   • B   • C

Circle 2

X •

• Y

A •   • B

Circle 3

A •   • B

• E

How Many Squares?

Make five squares using toothpicks as shown.

Move only two toothpicks to reduce the number of squares from five to four.

Toothpicks cannot overlap or be removed.

Make seven squares using toothpicks as shown.

Move only three toothpicks to reduce the number of squares from seven to five.

Toothpicks cannot overlap or be removed.

Four Cubes

Eight completely different objects can be made using four cubes. Each cube must be joined to another cube along at least one face.

Turning, flipping, or sliding a model does not make a new model. The four models below are only one answer.

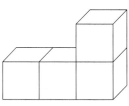
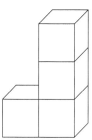
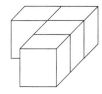
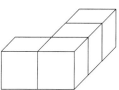

Draw all eight possible answers in the spaces below.

Octagon and Rectangle

Cut out the shapes below and rearrange them to make:

    (i)   an octagon

    (ii) a rectangle

Dividing Shapes

Divide the shape below into four.

Each quarter should be the same shape and size.

Divide the shape below into four.

Each quarter should be the same shape and size.

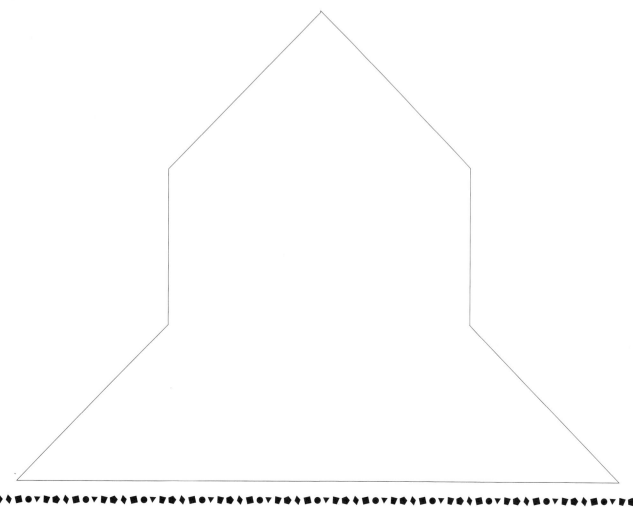

Lines and Crosses

Place nine crosses on the grid so that you can draw eight straight lines that each join three crosses.

Rearrange the nine crosses on the grid so that you can draw nine straight lines that each join three crosses.

Cube Models

How many cubes did it take to make each of the models below?

Model One: _____ cubes

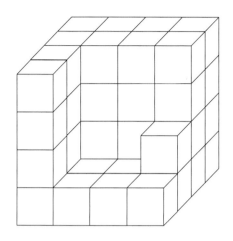

Model Two: _____ cubes

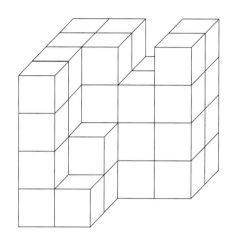

Model Three: _____ cubes

Model Four: _____ cubes

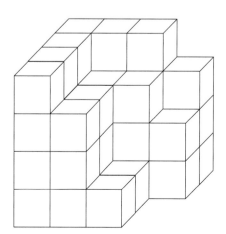

Cut and Make

Cut out the shapes below and rearrange them to make:

(i) a rectangle;

(ii) a square;

(iii) a triangle; and

(iv) a parallelogram.

Answers

Page 6: Straight Lines
Teacher check

Page 7: Curved Lines

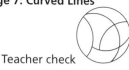

Teacher check

Page 8: Straight Line Pattern

Teacher check

Page 9: Rotating Squares

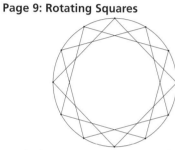

dodecagon – a polygon with twelve sides

Page 10: Parallel Lines

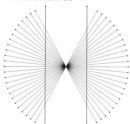

AB and CD appear to bend.

Page 11: Parallel Lines Again

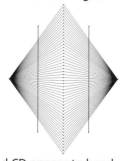

AB and CD appear to bend.

Page 12: The Pentagon
A pentagon has <u>5</u> sides and <u>5</u> corners. It has <u>5</u> diagonals.

Each side of the pentagon building would be ~320 m.

Page 13: The Hexagon
A hexagon has <u>6</u> sides and <u>6</u> corners. It has <u>8</u> diagonals.

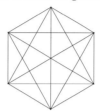

Jim lives in the <u>North</u> of Hexagon Island.

Page 14: The Octagon
Octagon Island has <u>8</u> sides. It has <u>8</u> corners. It has <u>20</u> diagonals.

1. S, 2. E, 3. N, 4. W,
5. NE, 6. NW, 7. SW, 8. SE

Page 15: The Nonagon

A nonagon has <u>9</u> sides and <u>9</u> corners. It has <u>27</u> diagonals.

nonagenarian – a person aged between 90 and 100 years
nonet – a composition for nine voices or instruments

Page 16: The Decagon

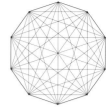

A decagon has <u>10</u> sides and <u>10</u> corners. It has <u>35</u> diagonals.

decade – a period of ten years
decathlon – an athletic contest made of ten different events
Decalogue – the ten commandments
decimal – pertaining to tenths of the number ten
December – the twelfth month of the year. Was once the tenth month of the old Roman year
decimate – to select by lot and kill every tenth man

Page 17: The Dodecagon

A dodecagon has <u>12</u> sides and <u>12</u> corners. It has <u>54</u> diagonals.

Page 18: Bunch of Squares!

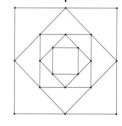

1. 14 cm **2.** 14 cm **3.** 14 cm
4. 14 cm **5.** a square **6. (a)** 14 cm
6. (b) 7 cm **6. (c)** 3.5 cm
Each length is half the previous length.

Answers

Page 19: Straight Curves

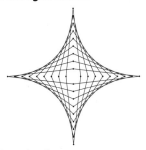

Teacher check

Page 20: Numbers and Letters

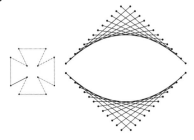

Page 21: Tessellation
Teacher check

Page 22: Compass Fun One

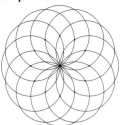

Page 23: Compass Fun Two

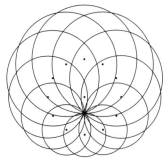

circus – in Roman times this was a circular arena surrounded by tiers of seats

circulates – to move or pass through a circuit back to the starting point, as in, from the heart, around the body and back to the heart

Page 24: Compass Fun Three

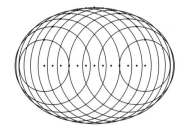

pile, lip, lips, sill, sell, pill, peel, slip, pie, ill, seep, sip, pies, pills, peels …

Page 25: Construct-a-Shield

Teacher check for shield design.

Page 26: Geometry and Heraldry

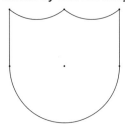

Teacher check for shield design.

Page 27: Instant 360° Protractor
Protractor construction sheet

Page 28: Instant Protractor Fun
1 to 6: teacher check

What sort of angle?
A. obtuse B. straight C. obtuse
D. reflex E. right F. right
G. right H. right I. reflex
J. obtuse K. reflex L. acute
M. reflex N. obtuse O. acute
P. reflex Q. acute

Page 29: More Instant Protractor Fun
Polygon construction: teacher check

sides	multiplied by =	number of diagonals
4	$\frac{1}{2}$	2
5	1	5
6	$1\frac{1}{2}$	9
7	2	14
8	$2\frac{1}{2}$	20

sides	multiplied by =	number of diagonals
9	3	27
10	$3\frac{1}{2}$	35
11	4	44
12	$4\frac{1}{2}$	54

Page 30: Tangram Fun
Tangram construction sheet

Page 31: More Tangram Fun

Page 32: Tessellation Again
Teacher check

Page 33: Flower Power

Other activities: teacher check

Page 34: Compass Cross

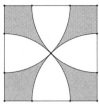

Answers

Page 35: Morning Star

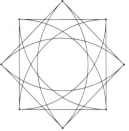

Page 36: Concentric Circles

Page 37: Parts of a Circle

Page 38: How Many Squares?

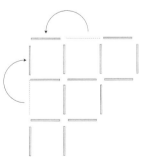

Page 38 (cont.): How Many Squares

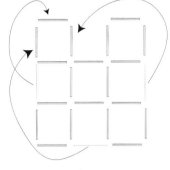

Page 39: Four Cubes

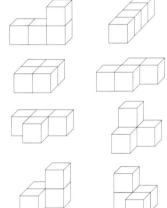

Page 40: Octagon and Rectangle

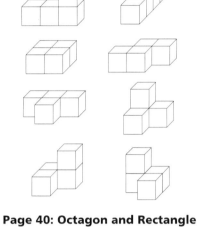

Page 41: Dividing Shapes

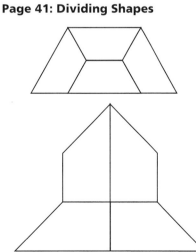

Page 42: Lines and Crosses

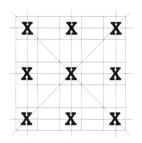

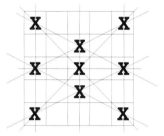

Page 43: Cube Models

Model 1: 47
Model 2: 41
Model 3: 48
Model 4: 43

Page 44: Cut and Make

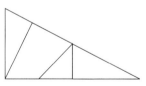

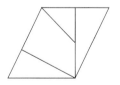